Luis Manuel Martínez
Valeria Elizabeth Fernandez
Orlando de la Osa Gonzalez

# Hot-melt adhesives with vegetable resins

Luis Manuel Martínez
Valeria Elizabeth Fernandez
Orlando de la Osa Gonzalez

# Hot-melt adhesives with vegetable resins

## Formulation and characterization of EVA/Grindelia mixtures

ScienciaScripts

**Imprint**
Any brand names and product names mentioned in this book are subject to trademark, brand or patent protection and are trademarks or registered trademarks of their respective holders. The use of brand names, product names, common names, trade names, product descriptions etc. even without a particular marking in this work is in no way to be construed to mean that such names may be regarded as unrestricted in respect of trademark and brand protection legislation and could thus be used by anyone.

Cover image: www.ingimage.com

This book is a translation from the original published under ISBN 978-620-3-03388-5.

Publisher:
Sciencia Scripts
is a trademark of
International Book Market Service Ltd., member of OmniScriptum Publishing Group
17 Meldrum Street, Beau Bassin 71504, Mauritius
Printed at: see last page
**ISBN: 978-620-3-24676-6**

# TABLE OF CONTENTS

# ABBREVIATIONS

Hc: Crystallization entalpy.

Hf: Fusion Entalpy.

DSC: Differential Scanning Calorimetry

E220: Ethylene vinyl acetate copolymer Elvax 220 W from Dupont,
Argentina.

E260: Ethylene vinyl acetate copolymer Elvax 260 from Dupont,
Argentina.

EVA: Ethylene Vinyl Acetate.

FTIR: Fourier Transform Infrared Spectroscopy

MI: fluidity index or softening point

PSA: pressure sensitive adhesives.

A: Rosin.

SBS: styrene-butadiene-styrene copolymers

SIS: styrene-isoprene-styrene copolymers.

Tc: Crystallization temperature.

Tg: Glass transition temperature.

Tgc: Glass transition temperature in cooling.

Tgm: Temperature of glass transition in heating.

Tm: Melting temperature.

TGA: Thermogravimetric Analyzer

VA: vinyl acetate.

# SUMMARY

In the present work, it was studied the effect produced when adding different amounts of a natural tackifier of vegetable origin (resin extracted from *Grindelia* genus) to Ethylene Vinyl Acetate (EVA) copolymers, commercially known as Elvax 220W and Elvax 260 (*DuPontTM*), of equal amount of Vinyl Acetate (28%p/p) and different Flow Rate (150 g/10min and 6 g/10min, respectively). The adhesives were formulated and prepared at laboratory scale. Then, the rheological, thermal and adhesive properties of the adhesives formulated with the two ethylene vinyl acetate copolymers as base polymer and the natural resins of the genus *Grindelia* as tackifier, as well as the physical, rheological and thermal properties of their components separately, were determined and analyzed. By Differential Scanning Calorimetry (DSC), in all the EVA/Tackifier mixtures, a good compatibility and miscibility of the systems studied was found due to the existence of a single glass transition temperature (Tg). The mixtures showed higher crystallization and melting values of ΔH than the values of the pure EVA copolymers, indicating that the tackifiers may have acted as a nucleating agent of the crystalline part of both copolymers.

Flow properties showed that all consistency index values decreased with increasing temperature for Grindelia resins. On the other hand, when analyzing the storage and loss modules (G' and G"), it was determined that the crossing temperature decreased as the percentage of tackifier increased for all the cases studied, evidencing that the tackifier caused, from the viscoelastic point of view, a temperature shift in the passage from elastic solid to viscous liquid. The shear resistance properties decreased when the tackifiers of the reference mixtures (EVA-Rosin) were replaced by the *Grindelia* resins. Nevertheless, they can be used as a replacement but for applications where the shear properties are not so demanding. From the results obtained from the characterization of the EVA/tackifier mixtures, the formulated adhesives of the hot-melt type are considered to be able to be used in the packaging industry considering the Grindelia resins as a replacement for pine resins extracted from heels from the wood industry.

# Chapter I

# Introduction General

# "Formulation and characterization of adhesive mixtures based on EVA (*Ethylene Vinyl Acetate*) modified with Grindelia resin"

## 1.1.  GENERAL INTRODUCTION

Adhesives are materials that are applied to the surface of two substrates or adhesives, and by means of bonding forces allow them to be bonded together in a manner that is resistant to separation. They form a bonding bridge at the interface of substrates, whether they are made of the same material or not, flexible and/or rigid. There are two phenomena that occur when trying to join two substrates, adhesion and cohesion. Adhesion is the set of physical and chemical interactions that take place at the adhesive/adhesive interface, while cohesion is the intermolecular forces of attraction, van der Waals forces, and the bonding forces that prevail between the molecules within the adhesive (Figure 1).

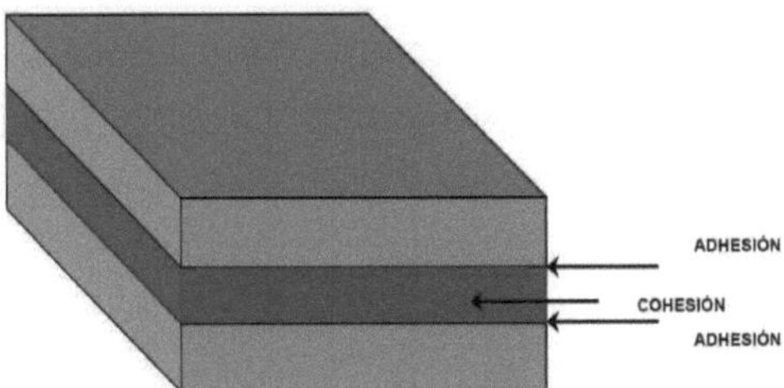


**Figure 1. Comparison between adhesion and cohesion in an adhesive joint. (Adapted from Loctite Spain)**

Adhesion between two substrates can be evaluated by performing an adhesive bond breakage test. If the failure occurs by adhesive breakage, the separation is said to have occurred by cohesion, if the separation occurs at the substrate/adhesive interface, it is referred to as separation by adhesion, or substrate breakage may occur directly before those mentioned (Figure 2).

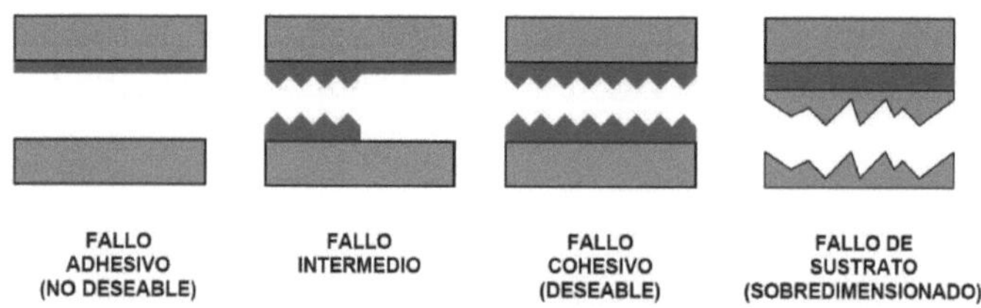



**Figure 2. Failure modes in an adhesive joint. (Adapted from Loctite Spain)**

There are two large groups of adhesives, pre-polymerized adhesives, those whose base polymer already exists within the adhesive itself before being applied to the materials to be joined, and reactive adhesives, characterized by being in a liquid, viscous, gel state, etc. They are made up of monomers or oligomeric chains that polymerize and/or cross-link during the polymerization process that occurs when such adhesive is located between the substrates to be joined.

Pre-curing adhesives exist in liquid or solid phase. Hot-melt adhesives are pre-polymerized adhesives that wet the substrates when heated above their softening temperature. When they cool, they acquire, quickly and without changes in their structure, the consistency of a plastic. In other words, they are 100% solid thermoplastic compounds at room temperature and liquid when applied, obtaining joints of high cohesion and very resistant to separation.

## 1.2.  POLYMERS

There is a great variety of polymers, resins and waxes that can be used in the manufacture of hot-melt adhesives for the packaging sector, among them are: ethylene vinyl acetate (EVA), polyamides, polyurethanes, polyester, etc. and are presented as viscous compounds, suitable for porous substrates such as paper, cardboard, fabrics, etc.

Hot melt adhesives are mixtures of three or more ingredients, so compatibility or miscibility between the components of the adhesive mixture is essential in order to relate it to the visco-elastic, thermal and adhesive properties of these adhesives. The most widely used base polymer for the formulation of hot melt adhesives, intended for the packaging sector, among others, is EVA copolymer.

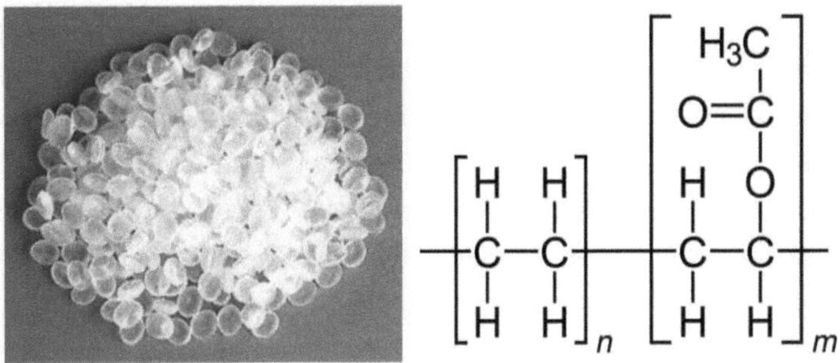

**Figure 3. Form of presentation and molecular structure of EVA (Wikipedia.org)**

These copolymers are obtained by copolymerization of ethylene and vinyl acetate (VA). The domain structure of EVA copolymers consists of rigid and partially crystalline polyethylene blocks of non-polar characteristics, and flexible polar and amorphous blocks of vinyl acetate (Figure 3). Their properties depend mainly on the molecular weight of the polymer and the vinyl acetate content in the structure. EVA copolymers are characterized by the % VA and the melt index (MI). The MI or melt point means the flow characteristic of hot-melt materials at fixed pressure and temperature, a parameter related to the molecular weight of the polymer. At higher molecular weights the polymer chains are longer, decreasing their ability to flow at fixed pressure and temperature so that the MI values will be lower than for those where the molecular weights are decreased. The higher the molecular weight, the higher the polarity and the lower the crystallinity. The variation in VA content also modifies the crystallinity of the copolymers. As it increases, crystallinity decreases, resulting in improvements in optical properties, flexibility and adhesion by reducing the softening point. EVA copolymers have a wide MI range, good adhesion properties to various adhesives and a low cost. Generally, the most commonly used EVA copolymers for the preparation of hot-melt adhesives have a vinyl acetate content between 18 and 28% w/w, although the most widely used is 28% w/w VA. One way to obtain EVA-based adhesives with a wide spectrum of properties is by combining this polymer with additives.

## 1.3.  TACKIFIERS

Among the most common additives used are fillers, tackifiers, antioxidants, among others. The amorphous tackifier resins of low molecular weight, play a fundamental role in hot-melt adhesives whose base polymer is a copolymer of ethylene vinyl acetate, since the copolymer alone does not present adequate adhesive properties. Therefore, tackifiers are commonly added to impart tack to hot-melt type adhesives and also in PSAs (pressure sensitive adhesives), formulated based on EVA copolymers and styrene-butadiene-styrene rubber copolymers (SBS) or styrene-isoprene-styrene (SIS). Typically, about 10-30% is added to commercial hot melt adhesives, providing desired viscosity without substantial decrease in adhesive bond strength or service temperature, improving processability, reducing cost and increasing flexibility of these adhesives. In addition, it must be reasonably compatible with the base polymer, have a very low molecular weight and have a glass transition temperature (Tg) that is higher than that of the base polymer. Therefore, the addition of sticking agents modifies the rheological and adhesion properties. The viscoelastic properties of EVA-based hot-melt adhesives were studied by different researchers (Marin, 1991; Marin, 1992; Vandermaesen, 1993; Tse, 1995), who proposed several theoretical models. Also, the compatibility of the sticking agent in mixtures with EVA were described and it was established that very polar tackifiers were more compatible with the vinyl acetate (VA) (polar) groups in the EVA copolymer, and that less polar and aliphatic sticking agents were more compatible with the ethylene (non-polar) groups of the EVA copolymer. Therefore, the VA content in the EVA copolymer and the nature of the tackifier will determine the compatibility, viscoelastic properties, and adhesives of the EVA/tackifier blends. The tackifying resins that can be incorporated into hot-melt adhesives can be classified into phenolic resins, hydrocarbon resins, coumarone-indene resins, and rosin resins and their derivatives. Rosin resin is one of the oldest resins used as a raw material for the adhesive industry, either as such (unmodified rosin) or in its rosin ester form (Figure 4). This resin is a solid resinous material and is obtained from turpentine or from the wood of pine trunks. It contains mainly an acid fraction, approximately 90%, and the remaining 10% are non-acid compounds. The neutral fraction of the rosin is made up of esters of resin and fatty acids. The acid fraction is a mixture of organic acids (abiotic and primary acids).

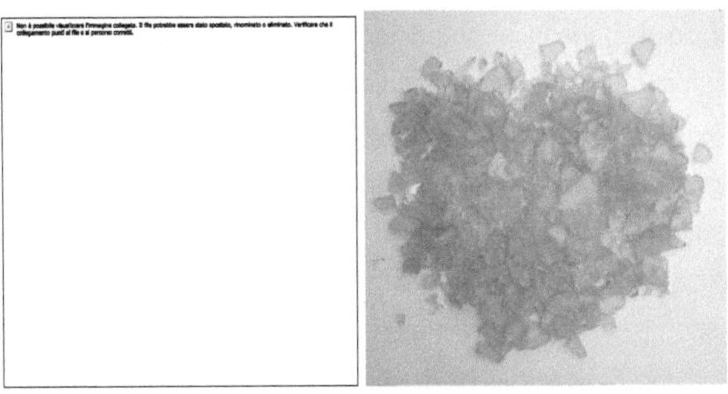



Figure 4. Industrial rosin resin presentation

The *Grindelia* Willd (Asteraceae) is an American genus comprising about 63 species. Hoffman (1986, 1984) selected Grindelia *comporum Green* as a promising species for the southwestern United States. Its resin, composed of diterpenic acids of the Labdan group, could replace those of pine (*Pinus* sp.) (Loeber, 1990). Since 1992, *Grindelia Chiloensis has* been planted in the Argentinean Patagonia in dryland conditions, with the objective of finding new productive options (Golluscio, 1999) to replace the agriculture and livestock that led to land degradation (Ravetta, 2002). This species, native to Patagonia, can yield up to 25% of terpene resins extracted from the leaves. It also produces resin on the surface of flowers and stems. Grindelic acid and diterpenic labdano acids make up 65% to 75% of the raw resin extracted from the plant. The antecedents that exist in Argentina with respect to the resin produced by the genus *Grindelia* refer mainly to the study of growth optimization and development of the cultivation, taking into account as a result the improvement in the production of biomass and resin (Wassner, 2002, 2005) (Figure 5).

Figure 5. Grindelia *Chiloensis* in Argentinean Patagonia

There are works on the chemical characterization of the resin composition of several *Grindelia* species, focused on the separation and structural identification by IR and NMR of some diterpenic acid species found in the resin (Guerreiro *et. al.*, 1981; Hanson, 2001; Ahmed *et. al.* , 2001). Other works studied the relationship between the solubility of these diterpenic acids with their toxicity and how it is reduced (Patoine, 1997; Peng, 2000). To achieve the application of the resin extracted from *Grindelia Chiloensis* as tackifier (adhesion modifier), given its high stickiness, studies have been carried out, mainly to know its flow behavior. Those that have shown that the refining process increases the viscosity of the resin, because the extraction of waxes produces an increase in the cohesive forces between the different components of the resin extracted from Grindelia (Martinez *et. al.*, 2009).

The extraction and purification of Grindelia resin has been done in solvents of different polarities, such as dichloromethane, ethanol and methane. The main difference between them is that the alcohols do not extract the wax from the plant, which has a possible use in the coating of fruits, given its antimicrobial characteristics. From the works of extraction and chemical characterization, it can be concluded that the Grindelia resin is constituted by diverse species that altogether present a wide miscibility, since it is soluble in solvents with different polarity indexes, such as 9.1 and 33 for dichloromethane and methanol respectively.

This makes the compatibility with the EVA copolymer adequate and enables the distribution of the Grindelia resin among the EVA chains. Table 1 shows some of the physical and thermal properties reported for the commercial EVA copolymers (Elvax® 220W and Elvax® 260) and Rosin used for the study of this work (Figure 6).

| Sample | Melt index (g/min) ASTM D1238 | Vinyl Acetate (%) | Density at 23ºC (g/cm3) ASTM D792 | Tensile Strength (MPa) ASTM D1708 | Elongation at Break (%) ASTM D1708 | Elastic Module (MPa) ASTM D1708 | Melting Point (ºC) DSC | Tg (ºC) DSC | Ring & Ball Softening Point (ºC) ASTM E28 |
|---|---|---|---|---|---|---|---|---|---|
| Elvax® 220W | 150 | 28,0 | 0,951 | 5,5 | 800-1000 | 16 | 70 | -35 | 88 |
| Elvax® 260 | 6,0 | 28,0 | 0,955 | 24 | 800-1000 | 26 | 75 | -30 | 154 |
| Rosin | - | - | - | - | - | - | - | 25 | 63 |

**Table 1. Typical physical and thermal properties of two EVA resins (Ethylene-Vinyl Acetate Copolymer) and unmodified rosin resin (Rosin)**

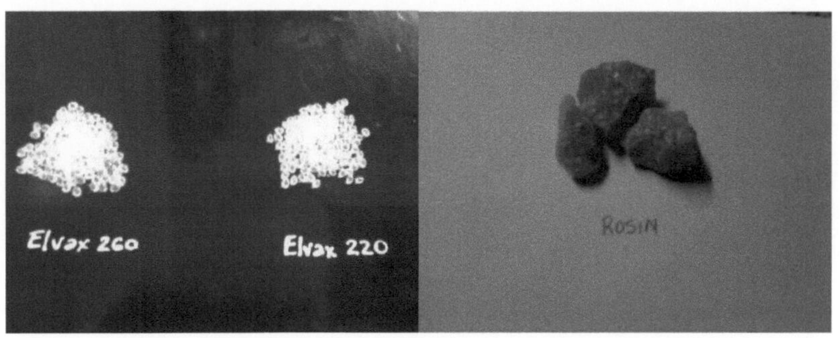


Figure 6. EVA copolymers (Elvax 260 and Elvax 220) and rosin resin.

# *Chapter II*

# *Materials*
# *Y*
# *Methods*

# MATERIALS AND METHODS

## MATERIALS

## 2.1.  OBTAINING THE GRINDELIA RESIN

The resins of the plants *Grindelia Chiloensis* and *Grindelia Camporum Greene,* were obtained from their leaves, stems and flowers in collaboration with Engineer Diego Wassner, member of the Faculty of Agronomy, National University of Buenos Aires.

### 2.1.1.  Raw resin extraction methodology

The leaves, stems and flowers, of both *Grindelia* species, were dried in an oven (model ULM 500, MEMMERT GmbH + Co.KG, Germany) at a temperature between 60 ºC and 65 ºC for 72 hours. After the drying period, they were ground using a manual grinder. 1 Kg of the crushed biomass was weighed, introduced into a 10 L ball and 3 L of dichloromethane (Anedra) was added. The ball was connected to a reflux cooling system and was brought to a boil for 8 hours. It was then cooled and vacuum filtered in a Buchner Kitasato to separate the rest of the biomass from the solvent-soluble resin. Finally, the solvent was evaporated by means of a Laborota 4000 rotavapor (Heidolph, Germany), thus obtaining the raw resin totally free of solvent (Figure 7).

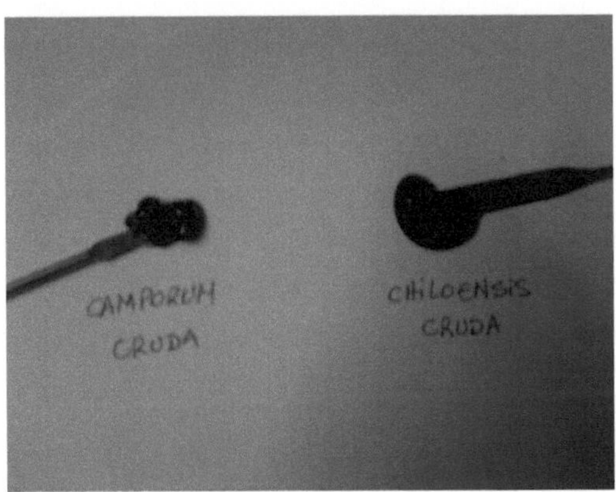


Figure 7. Raw resin of Grindelia *Camporum* and Grindelia *Chiloensis.*

### 2.1.2.  Raw resin refining

For the refining of the raw resin of *G. Chiloensis* and *G. Camporum Greene,* 300 g of each was introduced into a 1 L beaker and an equal volume of methanol (Anedra) was added. The mixture was treated at 50 ºC in a Masson model thermostatic bath (Vicking, Argentina), with constant agitation until phase separation was achieved. This generated the precipitate of the waxes contained in the raw resin. Once the system was cold, 100 ml more methanol was added and filtered by means of vacuum in a Buchner Kitasato, thus achieving the separation of the waxes from the resin dissolved in methanol. The absence of waxes was confirmed by adding a small volume of alcohol to this filtrate, without observing any precipitate. Activated carbon (Clarimex) was added to the filtrate, with a particle size of 75% ( 40 m), in a proportion of 7%, and was brought to a boil for 30 minutes with constant agitation in the same thermostatic bath. It was then filtered by means of vacuum, recirculating the filtrate until no remaining activated carbon was obtained on the filter paper. The remaining solvent was evaporated in a Laborota 4000 rotavapor (Heidolph, Germany), at a temperature between 60 ºC and 70 ºC, thus obtaining the refined resin of *G. Chiloensis* (Ch) and the solvent-free resin of *G. Camporum Greene* (C). Both resins were stored in glass flasks with lids for later use in the different tests carried out. A clear, dark brown resin was obtained for both *Grindelia* species.

## 2.2.  FORMULATION OF THE HOT-MELT ADHESIVES

In the formulation of the hot-melt adhesives, the copolymers Elvax 220W (E220) and Elvax 260 (E260) from the firm (DuPontTM) from Argentina were used, with the same amount of Vinyl Acetate (28% w/w) and different Flow Index (150 g/10min and 6 g/10min, respectively), as a polymeric base and the refined resins from *G. Chiloensis* and *G. Camporum Greene* as a stickiness modifying agent (tackifier). The unmodified rosin resin, Rosin WW (Brazilian origin) (R), was used as a reference tackifier. The formulations were carried out varying the EVA copolymers and the tackifier in the proportions indicated in Table 2.

| Sample | Composition (% w/w) | | | | | Sample Code |
|---|---|---|---|---|---|---|
| | Elvax 220W | Elvax 260 | *G. Camporum* | *G. Chiloensis* | Rosin | |
| Elvax 220W - *G. Camporum* | 90 | - | 10 | - | - | E220-C 90/10 |
| | 80 | - | 20 | - | - | E220-C 80/20 |
| | 70 | - | 30 | - | - | E220-C 70/30 |
| | 60 | - | 40 | - | - | E220-C 60/40 |
| | 50 | - | 50 | - | - | E220-C 50/50 |
| Elvax 220W- *G. Chiloensis* | 90 | - | - | 10 | - | E220-Ch 90/10 |
| | 80 | - | - | 20 | - | E220-Ch 80/20 |
| | 70 | - | - | 30 | - | E220-Ch 70/30 |
| | 60 | - | - | 40 | - | E220-Ch 60/40 |
| | 50 | - | - | 50 | - | E220-Ch 50/50 |
| Elvax 260 - *G. Camporum* | - | 50 | 50 | - | - | E260-C 50/50 |
| Elvax 260 - *G. Chiloensis* | - | 50 | - | 50 | - | E260-Ch 50/50 |
| Elvax 220W - Rosin | 50 | - | - | - | 50 | E220-R 50/50 |
| Elvax 260 - Rosin | - | 50 | - | - | 50 | E260-R 50/50 |

**Table 2. Composition of the formulated hot-melt adhesives**

## 2.3.  PREPARATION OF HOT-MELT ADHESIVES

For the preparation of the hot-melt adhesives, the corresponding EVA copolymer was introduced into a glass beaker. It was heated on a heating plate (Fbr Stir and Heat by Dealab SRL, Argentina) to a temperature between 160 ºC and 170 °C with constant agitation. Once the copolymer was melted, the refined resin of *Grindelia was* added, according to the adhesive that was formulated taking into account Table 2 and the temperature decreased to 100 ºC / 120 °C. In this way the mixture was thermally treated for 15 minutes maintaining constant agitation. Finally, the melted and homogeneous-looking mixture was poured onto a silicone plate and allowed to solidify at room temperature. Once solidified, the adhesive was separated from the silicone paper and stored inside a *nylon* bag in a desiccator, for later characterization.

## 2.4.  PREPARATION OF HOT-MELT ADHESIVE *FILMS*

The *hot-melt* adhesive *films* were obtained by compression molding at the Research Institute of Materials Science and Technology (INTEMA) of the National University of Mar del Plata.

A necessary amount of each EVA/tacker mixture, in its solid state, was used for the preparation of the *films.* On a hot plate press (EMS, Argentina), a mould 15 cm wide by 20 cm long and 0.3 mm thick was placed. The sample was placed inside the mold and heated to 95 ºC, under pressure, for 10 minutes using the press's own heating system. Once the mixture was melted, the pressure was increased for another 10 minutes with the piston, achieving the distribution of the mixture throughout the mold and the consequent formation of the *film.* To prevent the *film* from adhering to the press material, a *Teflon* film was used on both the upper and lower plates of the press. It was then cooled by recirculating water inside the press until it reached 25 ºC - 30 ºC. Each *film was then* removed and stored between separators in a desiccator for later characterization.

## METHODS

## 2.5.  CHARACTERIZATION OF THE *FILMS* OBTAINED

### 2.5.1.  Physical Properties

The physical properties that were measured to the hot-melt adhesives were density and characterization by Fourier Transform Infrared Spectroscopy (FTIR), in order to be able to select the substrates to evaluate their mechanical properties and thus determine their possible application.

### 2.5.1.1. Density

The density was determined for all hot-melt adhesives (Table 2) by the weight and volume of a 35 x 25 mm sample and the film thickness at 25 ºC. A digital balance (Peccisa Instruments AG, Stwizerland) with four decimal places and a manual digital gauge (Mitutoyo, U.S.A) were used for this purpose. Previously, the samples were conditioned in a desiccator for 24 hours.

### 2.5.1.2. Fourier Transform Infrared Spectroscopy Characterization

The characterization of the molecular structure of the crude and refined resin of *G. Chiloensis* and *G. Camporum* was determined by Fourier Transform Infrared Spectroscopy (FT-IR, *Fourier Transform InfraRed*) using an FT-IR Tensor 27 (Bruker, Spain) of resolution 1 cm-1, with an ATR of the Bruker Golden Gate Diamond type. The tests were carried out by determining the % of Transmittance as a function of wavelength. The characteristic bands of the main functional groups present in each sample were defined.

### 2.5.2.   Thermal Properties

Two different techniques were used for the thermal characterization of the samples, thermogravimetry and differential scanning calorimetry.

### 2.5.2.1.    Thermogravimetry (TGA)

The thermal properties of Elvax® 220W copolymer and the raw and refined resin of *G. Chiloensis* were characterized by thermogravimetry (TGA) using a TGA Q500 (TA Instrument, UK) (Figure 8). Samples ranging from 10 to 30 mg were taken. The test performed on each sample consisted of an initial heating at 5 ºC/min from 25 ºC to 100 ºC, and a second consecutive heating at 1 ºC/min from this temperature to 450 ºC. Considering that this last temperature was reached at constant weight, that is, there was no further loss of mass. Each test was performed under an atmosphere of N2 gas. This test was carried out with the objective of determining the temperature at which the components of the adhesives can be processed during the elaboration of the same.

The formulated adhesives corresponding to the EVA/tacker blends, E220-Ch 90/10, E220-Ch 80/20 and E220-Ch 60/40 were also characterized by TGA (Table 2). The influence of the degradation temperature of the individual components in the mixture obtained (EVA/tacker) is shown.

**Figure 8. Q500 Thermogravimetric Analyzer (TGA) (TA Instrument) with balance pan and heating oven detail.**

### 2.5.2.2. Differential Scanning Calorimetry (DSC)

The thermal properties of the hot-melt adhesives and their individual components were determined by *Differential Scanning Calorimeter (*DSC) using a Modulated DSC Q200 series (TA Instrument, UK), equipped with a Refrigerated Cooling System, RCS90 (TA Instrument, UK) (Figure 9). Prior to testing the MDSC was calibrated with Indium and Sapphire. Each sample of between 3 and 10 mg was placed in airtight aluminum capsules of type T zero. The test was performed by initial heating at 5 ºC/min from 25 ºC to 180 ºC to clear its thermal history. This was followed by cooling at 5 ºC/min to -85 ºC, and subsequent heating at the same speed to 180 ºC in the case of pure EVA samples. In the other samples, the same initial heating was performed, followed by a cooling and subsequent heating but in the range of -85 ºC to 120 ºC for rosin resin, -85 ºC to 100 ºC for *Grindelia* refined resins and -85 ºC to 110 ºC in the case of EVA/tackifier mixtures (Table 2). Each test was performed under an atmosphere of N2 gas at a constant flow rate of 50 ml/min.

From the curves obtained, the glass transition values (Tg) in cooling (Tgc) and heating (Tgm), the crystallization temperature and enthalpy (Tc and $_{\Delta Hc,}$ respectively) and the melting temperature and enthalpy (Tm and $_{\Delta Hf,}$ respectively) were determined.

**Figure 9. Differential Scanning Calorimeter (DSC) Q200 (TA Instrument) with RCS90 cooling unit**

## 2.5.3.  Rheological and Viscoelastic Properties

### 2.5.3. 1Flow Properties  (Rheology)

Rheological properties were determined using a Voltage Controlled Rheometer model AR-G2 (TA Instruments, UK) with a 40 mm diameter cone plate geometry and 2 ° cone angle. The separation between cone and plate *(gap) is* defined by the geometry used and corresponds to 55 m (Figure 10).

The flow behavior of the raw and refined resins of *G. Chiloensis and G. Camporum* were evaluated with the *flow test (stepped flow step)* between 60 ºC and 100 ºC controlling the temperature by means of a recirculating bath (Julabo model AWC 100) connected to the peltier plate of the rheometer. The range of shear deformation speed ($\gamma$The *shear rate* used was from 0.1 to 100 s-1 obtaining as response the *shear* stress ($\tau$Shear *stress*) in Pa.

### 2.5.3.2    Viscoelastic properties

The visco-elastic properties were determined using the same equipment used for rheological determinations as shown in Figure 10.

The storage modulus G' (Pa) and the loss modulus G" (Pa) of the EVA copolymers and the formulated hot-melt adhesives were obtained by means of a *temperature sweep* test between 25 ºC and 180 ºC, with the aim of determining the transition temperature from an elastic solid to a viscous liquid, by crossing these parameters.

**Figure 10. AR-G2 controlled voltage rheometer (TA Instrument) with cone plate geometry and recirculating bath (Julabo AWC 100) connected to the Peltier plate**

### 2.5.4.  Adhesive Properties

The adhesive properties of the formulated hot-melt adhesives were determined using a TC-500 Series II Universal Testing Machine (MegaTest, Argentina) (Figure 11). The shear strength test was performed, taking as reference the ASTM International D905 standard, from the *American Society for Testing* and Materials (ASTM), using a 500Kg load cell model CTC500 series nº 179 (MegaTest, Argentina).

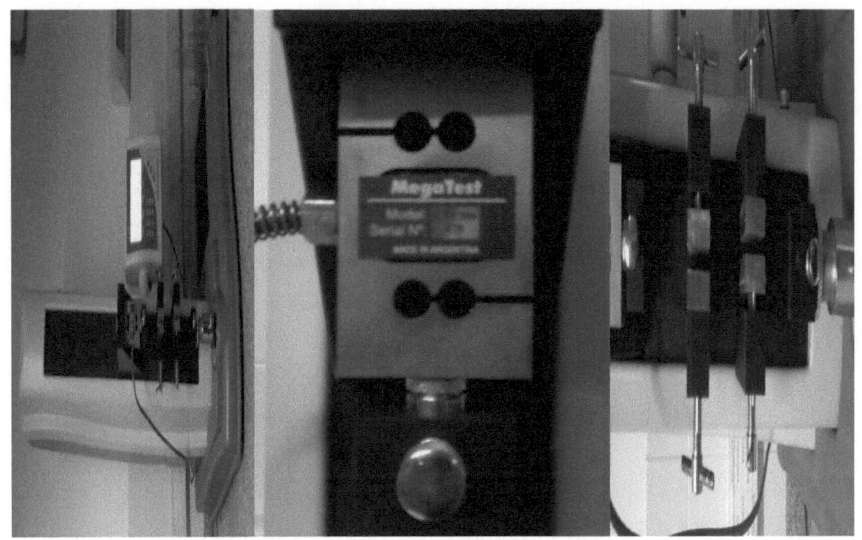

Figure 11. TC-500 Series II Universal Testing Machine (MegaTest, Argentina) with 500 Kg load cell model CTC-500 and manual clamps.

The adhesive joints were made at 25 ºC, with the adhesive in the form of a film preheating to 80 ºC. The film strip was placed on the first substrate (fiberglass reinforced plastic) and then the next substrate (fiberglass reinforced plastic) was placed on it and then pressed with a press for 10 minutes. The test was carried out 48 hours after the formation of the bond to the hot-melt adhesives indicated in Table 3.

| Sample | Composition (% w/w) | | | | |
|---|---|---|---|---|---|
| | Elvax 220W | Elvax 260 | *G. Camporum* | *G. Chiloensis* | Rosin |
| E220-C 50/50 | 50 | - | 50 | - | - |
| E220-Ch 50/50 | 50 | - | - | 50 | - |
| E260-C 50/50 | - | 50 | 50 | - | - |
| E260-Ch 50/50 | - | 50 | - | 50 | - |
| E220-R 50/50 | 50 | - | - | - | 50 |
| E260-R 50/50 | - | 50 | - | - | 50 |

Table 3. Samples for which the shear strength of adhesive joints was determined.

## 3.1 PHYSICAL PROPERTIES

### 3.1.1. Density

The determination of the density of the formulated adhesives showed values lower than 1 g/cm3, varying in the range of 0.850 - 0.976 g /cm3. This is valid considering that the density of the base polymers used is 0.951 g/cm3, for E220, and 0.955 g/cm3, for E260.

### 3.1.2. Structural characterization by Fourier transform infrared spectroscopy

From the FTIR spectra for the crude and refined resin of *G. Chiloensis* and *G. Camporum* (Fig. 12 and 13) it has been observed that both resins are structurally similar, although they come from different species of the genus *Grindelia.* This is in accordance with what has been reported in the literature, that both resins are chemically identical (Wassner and Ravetta, 2007).

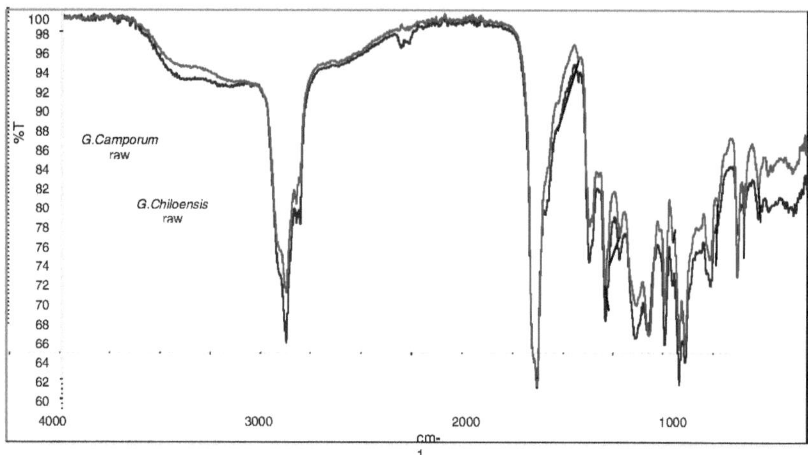


**Figure 12. FTIR spectra of the crude resins of *G. Chiloensis* (red) and *G. Camporum* (blue).**

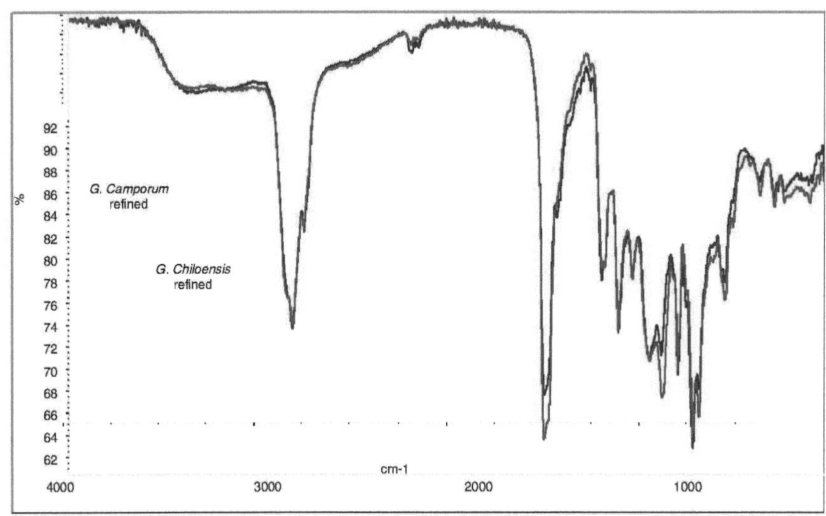

**Figure 13. FTIR spectra of the refined resins of *G. Chiloensis* (red) and *G. Camporum* (blue).**

In the FTIR spectra for the crude and refined resin of *G. Chiloensis* (Figure 14) the following bands have been characterized: in the region from 2950 to 2860 cm-1 the antisymmetric and symmetric bands of the methyl and methylene groups were observed, the bands in 1454 and 1440 cm-1 correspond to the CH2 deformation and in 1370-1380 cm-1 to the dimethyl group. In the region from 1225 to 988 cm-1 were found, for both resins, the characteristic bands of valence vibrations of the C-O- group of carboxyl. The main difference for them can be observed in the bands corresponding to the structure of the C=O, since the band of 1704 cm-1 of the carboxyl of the raw *Grindelia,* appeared in the refined *Grindelia* at 1732 cm-1. This implies that the process of refining into alcohol has had an effect on the band's slippage. On the other hand in the raw resin was observed an asymmetric doublet at 735 cm-1 and 701 cm-1 which is related to the structure of OH of the acid group, but can also be attributed to traces of olefinic waxes, since for the samples of refined resin that signal disappeared.

23

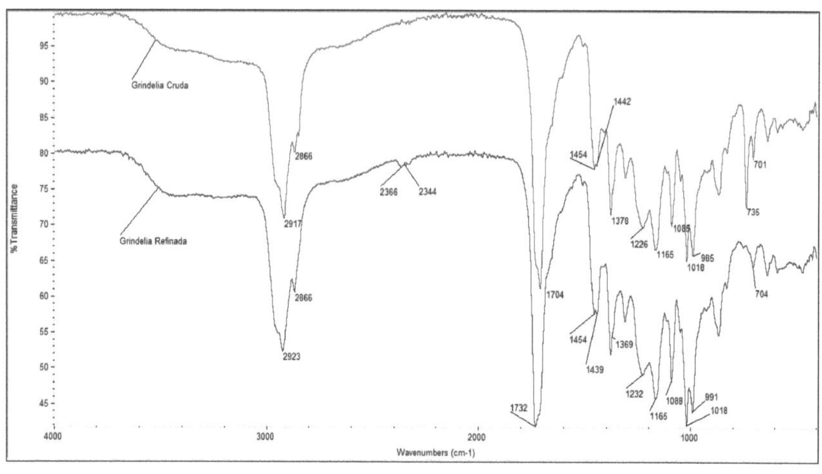


Figure 14. FTIR spectra of Grindelia *Chiloensis* resins.

## 3.2. THERMAL PROPERTIES

### 3.2.1. Thermogravimetry (TGA)

From the analysis of the thermograms obtained by TGA, it was observed that the E220 copolymer presented two well-defined regions of mass loss (Figure 15). The region with the lowest temperature, occurred close to 350 ºC and may correspond to the amorphous vinyl acetate fraction. The higher temperature region occurred at 450 ºC and may correspond to the ethylene fraction of the copolymer. The thermograms of *G. Chiloensis* (Figure 16) showed mass loss difference between its raw and refined resin around 100 ºC, which is attributed to the presence of volatile compounds in the raw resin. For both resins their thermal stability reached 200 ºC. This makes that when EVA copolymers are added when preparing the adhesives, the mixture loses thermal stability in relation to the pure EVA, appearing loss of mass from 200 ºC (Figures 17, 18 and 19), which becomes more sensitive as the content of *Grindelia in the* mixture increases. From the results obtained, it should be noted that the preparation temperature of EVA hot-melt adhesives and Grindelia resins should never exceed 200 ºC.

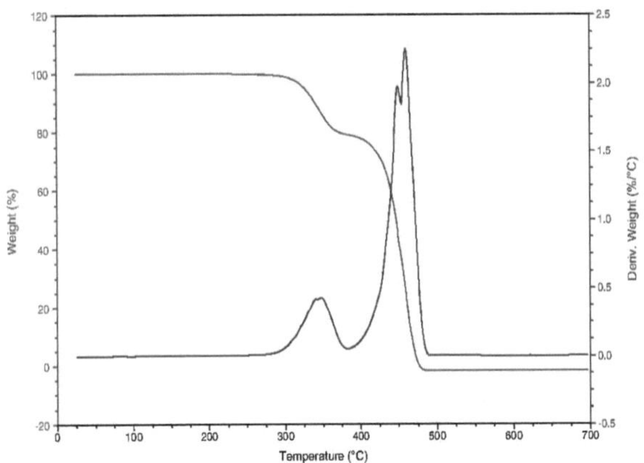

**Figure 15. Mass loss as a function of E220 copolymer temperature.**

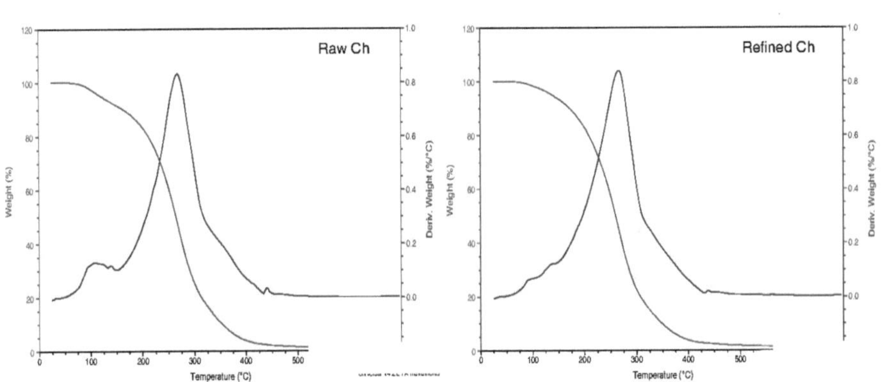

**Figure 16. Temperature-dependent mass loss of *G. Chiloensis* raw and refined resin.**

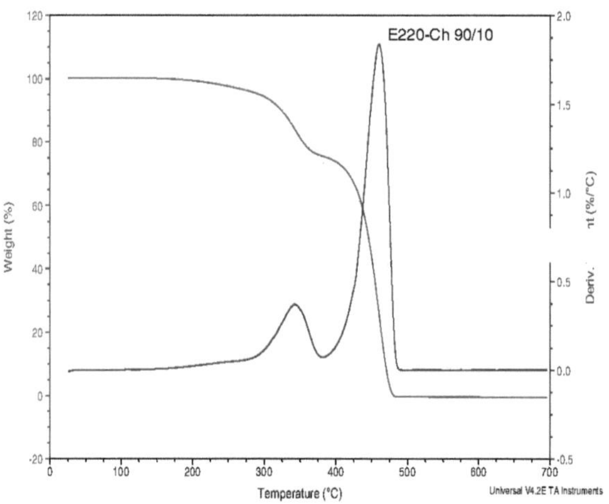


**Figure 17. Loss of mass as a function of adhesive temperature E200-Ch 90/10.**

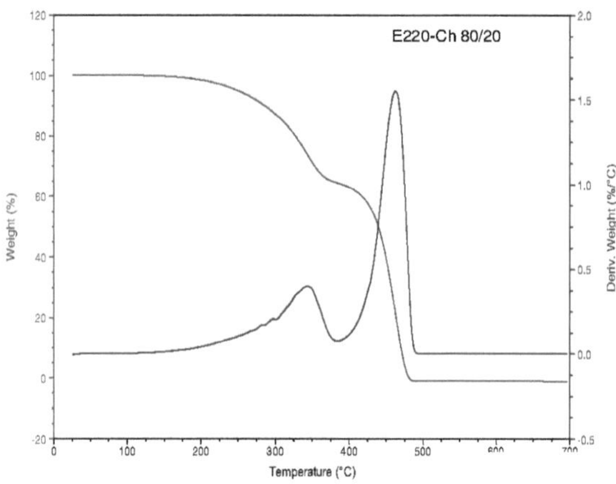


**Figure 18. Loss of mass as a function of the temperature of the adhesive E220-Ch 80/20.**

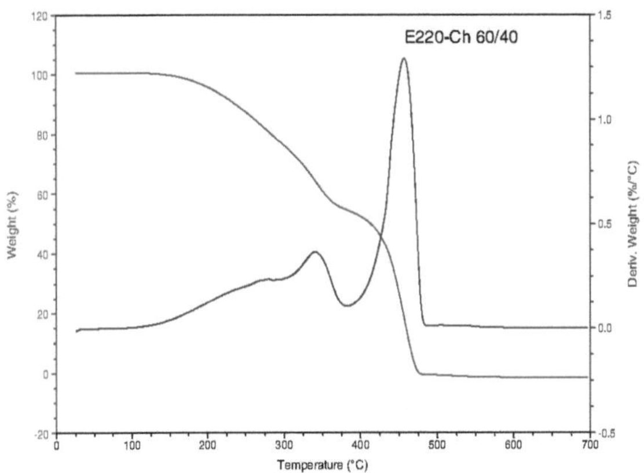

**Figure 19. Loss of mass as a function of adhesive temperature E220-Ch 60/40.**

### 3.2.2.    Differential Scanning Calorimetry (DSC)

Figure 20 shows the thermograms of the commercial EVA copolymers, E220 and E260, used in this work. In it, all the thermal transitions that occur in both ethylene vinyl acetate copolymers are observed qualitatively and quantitatively. The data calculated for Tg, Tm and Tc, both in the melting process and for crystallization, correspond to those reported (Table 1) in the technical sheets of the products. Tg was found between -30 °C and -35 °C and Tm around 70 °C. These values were found close to those reported in the technical sheets of these copolymers, 70 °C and 75 °C respectively, but not equal. It may have probably happened that the heating rate was not exactly 5 °C/min, as the experiments reported here were performed. Although it is important to highlight that the trend of Tm values were maintained, that is, the Tm of E220 copolymer was lower than that of E260. This same trend could also be observed in the Tc for both copolymers. The correspondence between Tc and Tm with MI is that as the MI decreases (150 g/10min for E220 and 6 g/10min for E260) Tm and Tc increase, due to the difference in their molecular weights from EVA copolymers (E220 and E260).

27

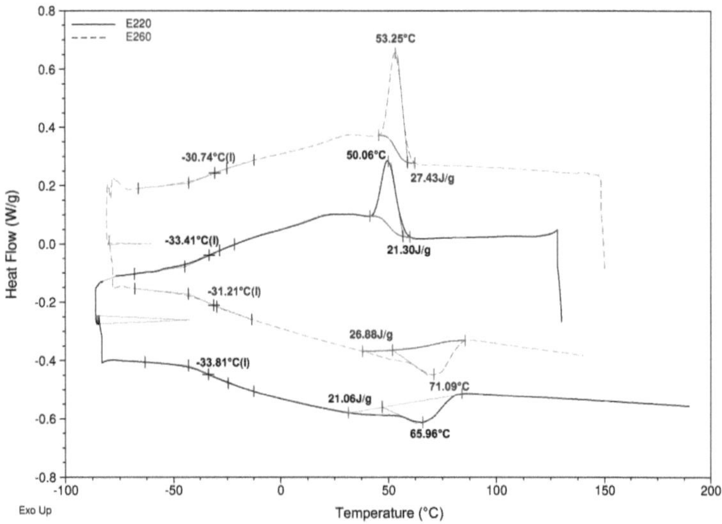

Figure 20. Heat flow vs. temperature of E220 and E260.

Figure 21 shows the thermogram with the experimental results obtained for the unmodified rosin resin (Rosin), where the Tg calculated both in cooling and heating is close to 25 ºC as reported by Aran Ais *et al*, 2000; Table 1.

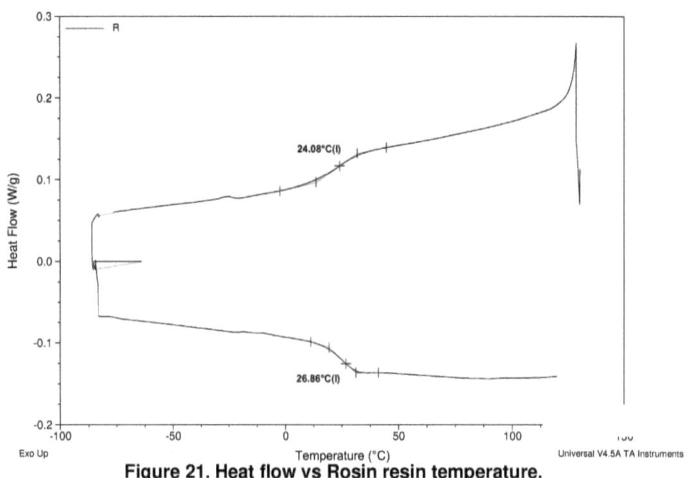

Figure 21. Heat flow vs Rosin resin temperature.

The thermal behavior of the refined resins of G. *Chiloensis* and G. *Camporum* is shown in Figure 22. Given the structural characteristics of both *Grindelia* and Rosin resins, they present only glass transition as a thermal phenomenon, clearly showing their amorphous state, typical of a natural resin. The Tg for G. *Camporum* is somewhat higher than for G. *Chiloensis*, which may be linked to a difference in the distribution of their composition, since they are extracted from silver of different species. Although not in terms of their constituents, as both resins were reported to be chemically identical (Wassner and Ravetta, 2007). In comparison with the rosin resin, used as a reference tackifier, the *Grindelia resins* present a lower Tg which will have as a result that in the mixtures with EVA a lower influence in the increase of the Tg of the mixture is observed.

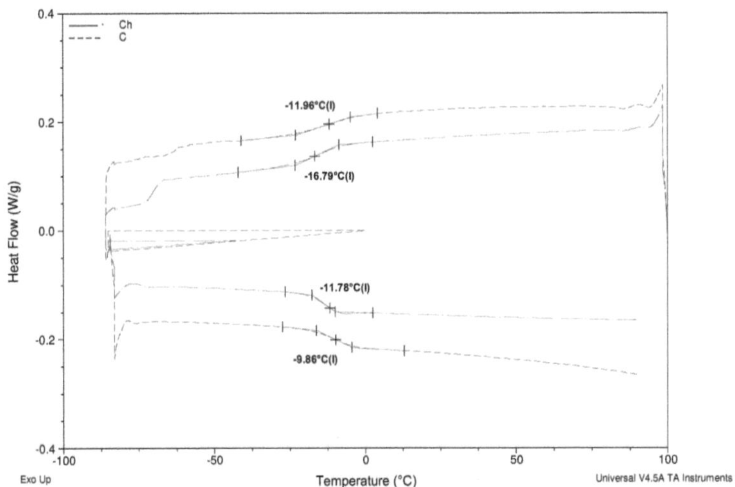


**Figure 22. Heat flow vs. temperature of G. *Chiloensis* and G. *Camporum* refined resins**


The thermographs of the mixtures of E220 and E260 with the different tackifiers (Ch and C refined resin and Rosin resin), in the maximum proportion used (50-50 % w/w), are shown in Figure 23 and 24. The main objective of comparing the thermal transitions of the EVA/tackifier mixtures in different proportions, was to study the degree of miscibility and compatibility that the components used in them present. Since these two behaviors are of primary importance for the formulation of an adhesive. As it can be seen, the mixtures of EVA copolymers with Rosin only presented one Tg, for the mixture with E220, the Tg was -16 ºC and -11 ºC, and for the mixture with E260, very close to -14 ºC, in the heating and cooling process.

29

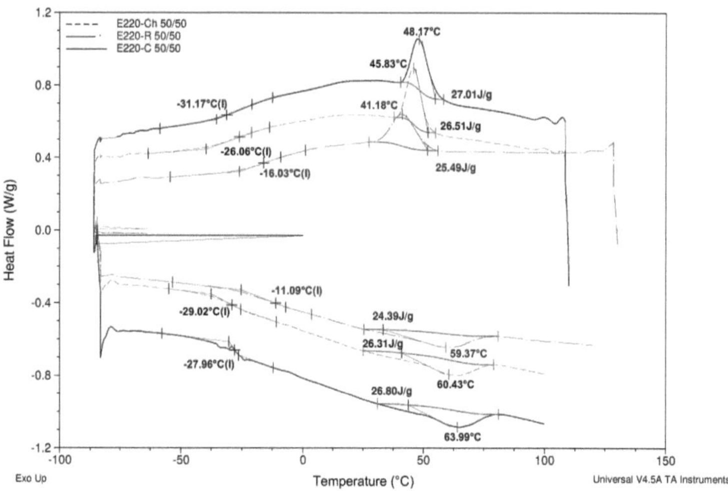

Figure 23. Heat flow vs. temperature of adhesives formulated with E220 and the refined resins of G. *Chiloensis* and G. *Camporum* and Rosin at 50% w/w.

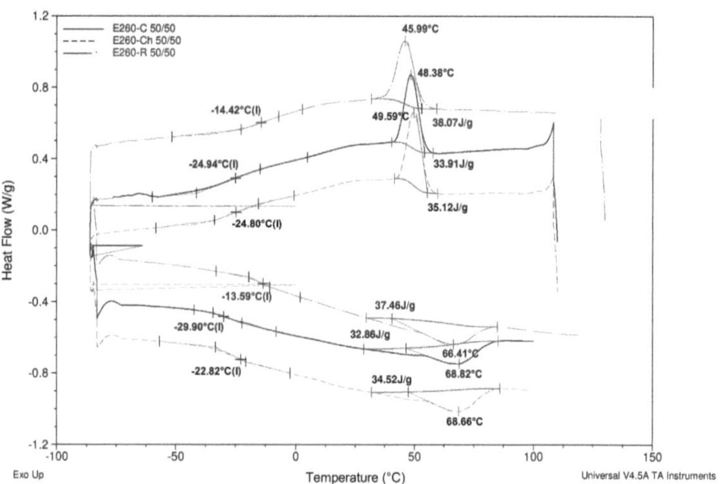

Figure 24. Heat flow vs. temperature of adhesives formulated with E260 and the refined resins of G. *Chiloensis* and G. *Camporum* and Rosin at 50% w/w.

On the basis of all the consulted bibliography this behavior, in which only one Tg is observed for the mixture and that the same one is between the Tg of the materials separately, without concerning that these materials by themselves present very different Tg, indicates a very good compatibility of the components (base miscibility between them. This could be appreciated during the preparation of the adhesives, since completely homogeneous mixtures were obtained, i.e. without any phase separation. This behavior, also observed in rosin resins, is one of the reasons why they are widely used as industrial tackifier with EVA in the formulation of adhesives.

Table 5 summarizes the thermal characteristics of the samples of interest for the analysis in its pure state and mixed with the tackifiers used in a proportion of 50-50 % w/w.

| Sample Code | Crystallization | | | Merger | | |
|---|---|---|---|---|---|---|
| | Tc (ºC) | ΔHc (J/g) | Tgc (ºC) | Tm (ºC) | ΔHm (J/g) | Tgm (ºC) |
| E220 | 50,06 | 21,30 | -33,41 | 65,96 | 21,06 | -33,81 |
| E260 | 53,25 | 27,43 | -30,74 | 71,09 | 26,88 | -31,21 |
| C | - | - | -11,96 | - | - | -9,86 |
| Ch | - | - | -16,79 | - | - | -11,78 |
| R | - | - | 24,08 | - | - | 26,86 |
| E220-C 50/50 | 48,17 | 27,01 | -31,17 | 63,99 | 26,80 | -27,96 |
| E220-Ch 50/50 | 45,83 | 26,51 | -26,06 | 60,43 | 26,31 | -29,02 |
| E260-C 50/50 | 48,38 | 33,91 | -24,94 | 68,82 | 32,86 | -29,90 |
| E260-Ch 50/50 | 49,59 | 35,12 | -24,80 | 68,66 | 34,52 | -22,82 |
| E220-R 50/50 | 41,18 | 25,49 | -16,03 | 59,37 | 24,39 | -11,09 |
| E260-R 50/50 | 45,99 | 38,07 | -14,42 | 66,41 | 37,46 | -13,59 |

**Table 5. Thermal properties of EVA copolymers, tackifiers (refined resins) and hot-melt adhesives formulated with 50% w/w mixtures of these components.**

Figure 23 and 24 also show the thermograms obtained for the 50-50 % w/w mixtures between the refined resins of G. *Chiloensis* and G. *Camporum* with the copolymers E220 and E260 respectively. In these, as with Rosin resin, it was possible to calculate only one Tg in both the cooling and heating stages. This leads us to conclude that also both *Grindelia* resins have a very good compatibility and miscibility with the EVA copolymers used, independently of their MI, or what can also be translated as different molecular weights of the copolymer. As for the Tg values found for these mixtures, it can be observed that they are lower than those obtained with Rosin resin. This is correct if we take into account that the refined *Grindelia* resins (Figure 22) present a lower Tg than the Tg of the rosin. Therefore, the Tg of the blends will look much closer to the Tg of the corresponding pure EVA copolymer. This technologically influences only the elastic properties of the adhesive formulated with the *Grindelia* resins. These properties will be maintained at lower temperatures than the properties of the adhesives formulated with Rosin resin as a tackifier. As for the Tc and Tm in all cases decrease with the presence of the tackifier.

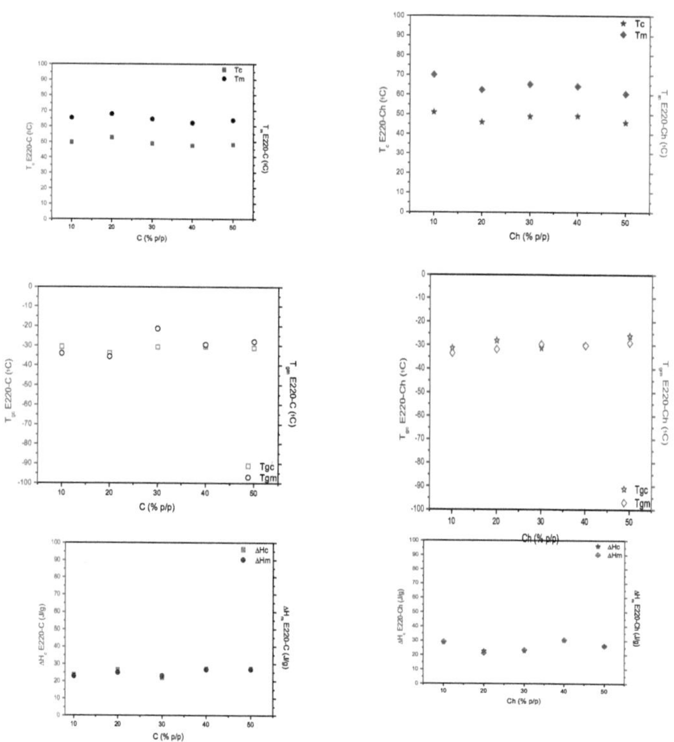

**Figure 25. Variation of thermal transitions (Tc, Tm, Tg and ΔH) during crystallization and melting of adhesives formulated with tackifier content (G. *Chiloensis,* G. *Camporum and* Rosin).**

However, Rosin resin makes this descent more pronounced than *Grindelia* resins. And in all cases also the tackifier apparently has an effect on the nucleation process during crystallization, since for the mixtures with E220 and E260 an increase in the crystallization ΔH and therefore in the melting process can be seen, compared to the values obtained for pure EVA copolymers. The variation of the *Grindelia* tackifier (10-50 % w/w) in the mixture with EVA, shows that its fluctuation only slightly affects the values of the thermal transitions (Tc, Tm, Tg and $\Delta$H) with respect to those of pure EVA. In other words, the effects on thermal properties are linked to the presence of the tackifier, but not to its quantity (Figure 25).

## 3.3.    RHEOLOGICAL AND VISCOELASTIC PROPERTIES

### 3.3.1.    Flow Properties (Rheology)

When analyzing the results obtained in the flow test, for the refined and crude resins of *G. Chiloensis* and G. *Camporum* (Figure 26 and 27), it was found that the best model that fitted the experimental points of each curve was the Herschel-Bulkley model. This model associates a threshold stress parameter ($\tau_o$ The following expressions are used to describe this *process*:

Power Law: $$\tau = K(\dot{\gamma})^n$$

Herschel-Bulkley model: $$\tau = \tau_o + K(\dot{\gamma})^n$$

where $\tau$ is the cutting effort, $K$ the consistency index, $\dot{\gamma}$ the speed of deformation by cutting, $n$ the flow behaviour index and $\tau_o$ the threshold effort.

To obtain their parameters, the experimental curves were adjusted by linear regression analysis, using the expression:

$$y = a + b * x^c \text{ where } a = \tau_o, \, b = K \text{ y } c = n$$

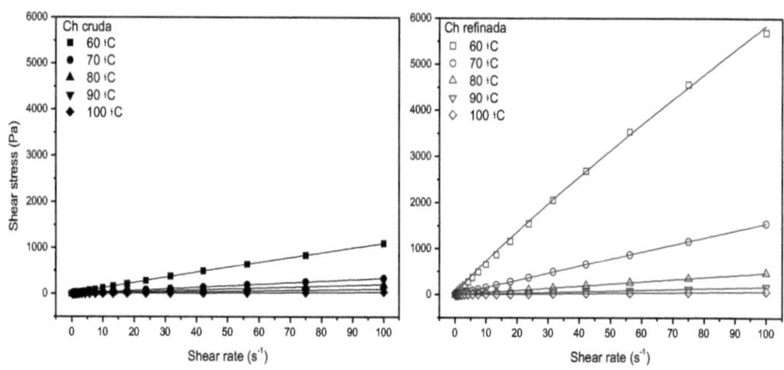

**Figure 26. Flow behavior of *G. Chiloensis* raw and refined resin.**

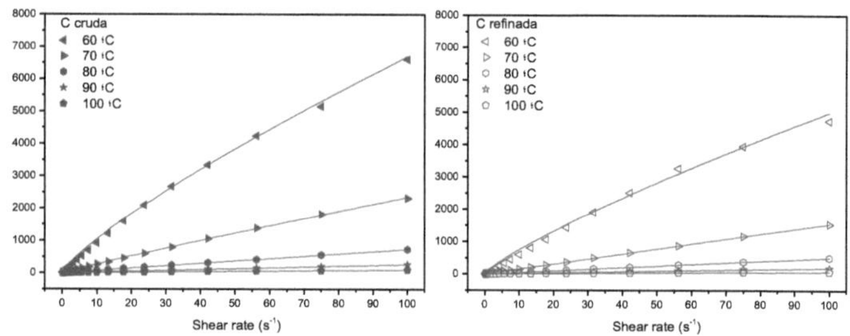

**Figure 27. Flow behavior of *G. Camporum* raw and refined resin.**

In Figure 26 you can see that the refined resin of G. *Chiloensis* increases its flow behavior with respect to the raw resin. The inverse behavior is shown in Figure 27 for the resin of *G. Camporum, that* is, the flow behavior of its raw resin is different from that of the refined one. This corresponds to the K values obtained in each case. The decrease in the K index (consistency), between 164 and 2 Pa s, depending on the temperature in the raw and refined resins would indicate a decrease in apparent viscosity to 80 ºC, and then remain constant between approximately 2 and 0.45 Pa s (Figure 28). In general, it can be said that as the temperature increases, the consistency value for *Grindelia* resins decreases until it remains almost constant in which the cohesion forces of both become similar. On the other hand, at the same temperature in the (Figure 28) it can be seen that the raw resin of *G. Chiloensis* turned out to be more viscous than its refined one (K $_{Ch\ raw}$ > K Ch $_{refined}$), this may be due to the interactions that occur between the components of the resin and the separated vegetable wax during the refining. The results show that in the absence of the wax, the cohesive forces between the components of the refined resin increase. In the temperature range of 60 ºC to 100 ºC, the fraction of wax and resin constitute a single phase, which implies a great interaction between all the components of the mixture, resulting in a higher apparent viscosity (Martinez, *et al.* 2009). On the other hand, *G. Camporum*'s crude resin turned out to have lower viscosity than its refined resin at the same temperature (K $_{C\ crude}$ > K C $_{refined}$) due to the natural absence of waxes in the former. Probably the extraction of high molecular weight components from the raw resin will make the resulting refined resin more viscous.

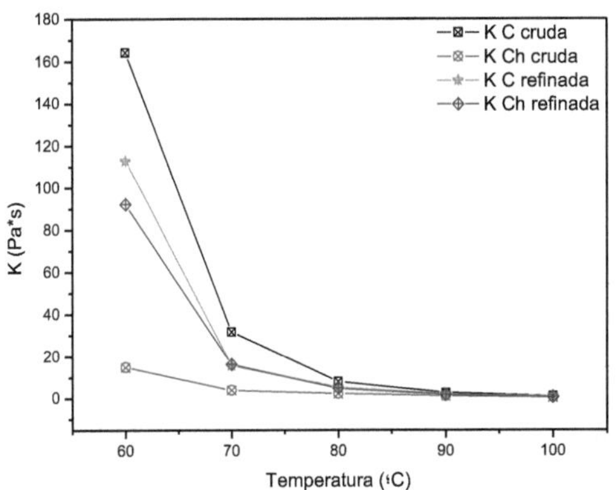


**Figure 28. Consistency index (K) of raw and refined resins of *G. Chiloensis* and *G. Camporum*.**

For the purpose of the results obtained by the adjustment of the Herschel-Bulkley model adopted, we can say that the resins presented a non-Newtonian fluid behavior with pseudoplastic character since the values of the parameter n were in all cases less than 1 (Table 6). Nevertheless the previous thing, as all the values of n are very near to 1 we can consider, practically, that the resins tend to a behavior of Newtonian fluid where K would be representing the value of the absolute viscosity of the crude and refined resins of *G. Chiloensis* and *G. Camporum*.

| Sample | Temperature | Parameter | | | |
|---|---|---|---|---|---|
| | | $_0$ (Pa) | (Pa.$^{sn}$) | n | R2 |
| Raw | 60ºC | 0,0000 | 164,28 | 0,8050 | 0,9991 |

| | | | | | |
|---|---|---|---|---|---|
| *Camporum* | 70ºC | 0,0000 | 31,81 | 0,9340 | 0,9997 |
| | 80ºC | 0,0000 | 7,96 | 0,9807 | 1,0000 |
| | 90ºC | 0,0620 | 2,67 | 0,9974 | 1,0000 |
| | 100ºC | 1,0516 | 0,98 | 0,9944 | 0,9984 |
| Raw *Chiloensis* | 60ºC | 0,0000 | 14,95 | 0,9340 | 1,0000 |
| | 70ºC | 0,0000 | 3,93 | 0,9671 | 1,0000 |
| | 80ºC | 0,0521 | 2,35 | 0,9704 | 1,0000 |
| | 90ºC | 0,0000 | 1,09 | 0,9907 | 1,0000 |
| | 100ºC | 0,1139 | 0,45 | 0,9425 | 0,9999 |
| Refined *Camporum* | 60ºC | 0,0000 | 113,00 | 0,8225 | 0,9961 |
| | 70ºC | 0,0000 | 15,77 | 0,9960 | 1,0000 |
| | 80ºC | 0,0000 | 5,12 | 0,9958 | 1,0000 |
| | 90ºC | 0,0000 | 1,93 | 0,9965 | 1,0000 |
| | 100ºC | 0,5696 | 0,89 | 0,9534 | 0,9991 |
| Refined *Chiloensis* | 60ºC | 0,0000 | 92,38 | 0,9003 | 0,9991 |
| | 70ºC | 0,0000 | 16,24 | 0,9901 | 1,0000 |
| | 80ºC | 0,0000 | 4,84 | 0,9941 | 1,0000 |
| | 90ºC | 0,0073 | 1,71 | 0,9998 | 1,0000 |
| | 100ºC | 0,2089 | 0,70 | 0,9979 | 0,9998 |

**Table 6. Rheological parameters as a function of temperature of raw and refined resins of *G. Chiloensis* and *G. Camporum*.**

### 3.3.2. Viscoelastic properties

The results obtained from the studies carried out are shown in Figures 29, 30, 31 and 32. The graphs show the variation of the storage module G' (Pa) and the loss module G" expressed in Pa of the EVA copolymers, (E220 and E260) and of the formulated hot-melt adhesives, as a function of the temperature. In Figure 29, the experimental curves of E220 and E260 are shown, in them it was seen that the temperature at which these copolymers pass from an elastic solid (G' > G") to have viscous liquid behavior (G' < G") from a certain point where it is fulfilled G' = G".

The value found where G' = G" is around 60 ºC (Table 7). This value is similar to the reported value (61.2 ºC) in the literature for 28% VA EVA copolymers (Barrueso-Martínez, *et al.* , 2001).

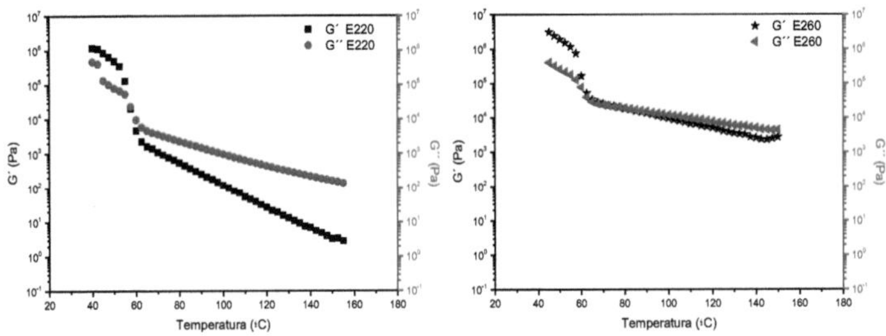

**Figure 29. Storage module G' and loss module G" of the E220 and E260 copolymers.**

The adhesives formulated with E220 and the refined resins of *G. Chiloensis* and *G. Camporum* in the proportions between 10 % w/w and 50 % w/w (Table 7), presented values of G' > G" up to the range of temperature between 50 ºC and 80 ºC, indicating that up to that point the behavior of elastic solid was dominant. Then the values of G" were higher than those of G', going on to behave as a viscous liquid. The increase of the tackifier content (G. *Chiloensis* and *G. Camporum)* produced an increase in the value of the crossing module (Gc = G' = G") and a decrease of the crossing temperature (Tcruce) due to the increase of the amorphous fraction and this accelerates the point in which the structural disorder of the mixture is produced (Table 7).

The adhesives at 50% w/w (Figures 30 and 31) of the EVA copolymers and *Grindelia* resins showed the same behavior but the temperature of crossing from elastic solid to viscous liquid was between 52 ºC and 58 ºC. For the adhesive formulated with EVA copolymers and Rosin resin at 50% w/w (Figure. 32), the behavior was of an elastic solid up to approximately 47 ºC in the case of E220 and 63 ºC in the adhesive with E260, and after these temperatures it became a viscous liquid.

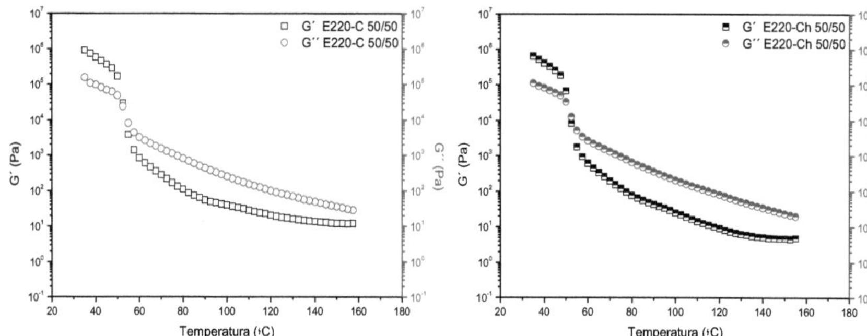

Figure 30. E220-G's 'G' storage module and 'G' loss module *Camporum* 50/50 and E220-G. *Chiloensis* 50/50.

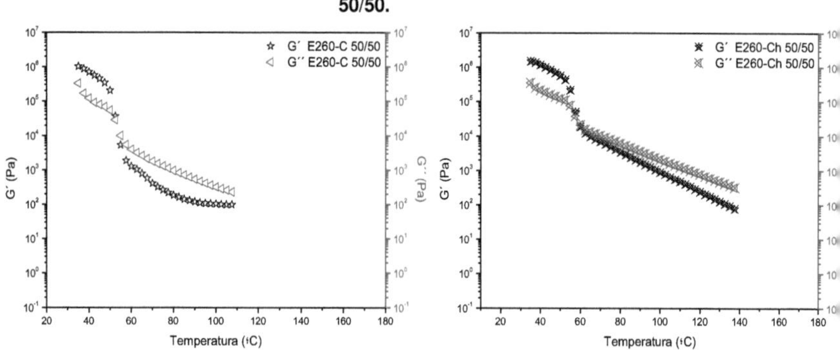

Figure 31. Storage module G' and loss module G'' of E260-G. *Camporum* 50/50 and E260-G. *Chiloensis* 50/50.

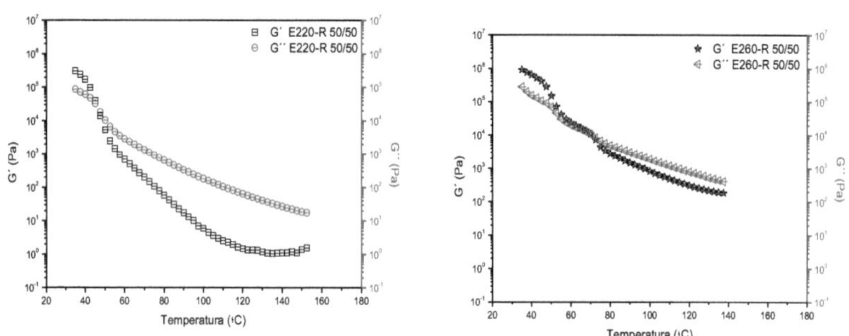

Figure 32. G' storage module and G' loss module of E220-Rosin 50/50 and E260-Rosin 50/50.

| Sample Code | Crossing temperature (ºC) | Gc (Pa) |
|---|---|---|
| E220 | 57,5 | 21660 |
| E260 | 62,5 | 45375 |
| E220-C 50/50 | 52,5 | 26960 |
| E220-C 70/30 | 55 | 18770 |
| E220-C 90/10 | 57,5 | 10057 |
| E220-Ch 50/50 | 50 | 50960 |
| E220-Ch 70/30 | 75 | 19635 |
| E220-Ch 90/10 | 77,5 | 1804 |
| E260-C 50/50 | 52,5 | 32335 |
| E260-Ch 50/50 | 57,5 | 44115 |
| E220-R 50/50 | 47,5 | 16160 |
| E260-R 50/50 | 62,5 | 18265 |

**Table 7. Temperature and cross G for EVA copolymers and adhesives formulated with the resin of *G. Camporum*, *G. Chiloensis* and *Rosin***

In general, it can be concluded that at low temperatures G' and G" acquired very high values up to a certain temperature (Tcruce), and then decreased sharply. G' and G" cross at a point corresponding to a change in the properties of the polymer. For temperatures below the Tcruce, the resin has primarily elastic properties, while for higher temperatures, the viscous behavior is more pronounced. Once the temperature at which the cut between the G' and G" modules is exceeded, the creep of the polymer chains occurs (Aran Ais, 2003) due to a significant structural disorder.

## 3.4. ADHESIVE PROPERTIES

### 3.4.1. Resistance to cutting

Figure 33 shows the maximum shear resistance values obtained when testing E220 and E260 adhesives with Rosin resin, used as a reference. It can be seen that the maximum resistance is lower in the case of the mixture with E220 than that obtained with E260, due to the greater molecular disorder that E220 presents, taking into account that the molecular weight of the latter is greater and therefore the packing of its crystalline fraction is lower. This is reflected in the MI, given that E220 has a higher MI and a higher molecular weight with respect to E260, whose MI is much lower and therefore has a lower molecular weight.

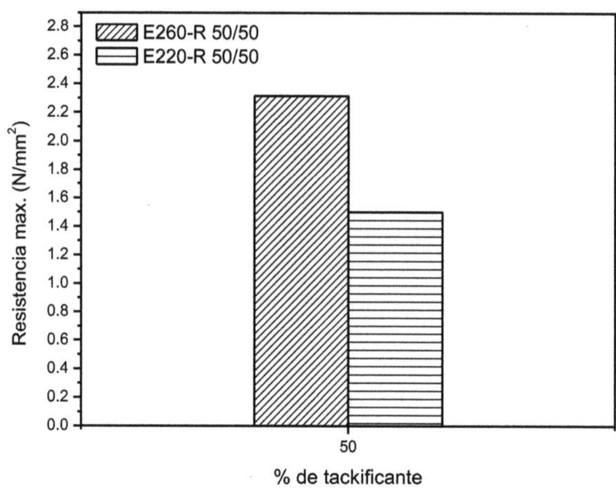

Figure 33. Maximum resistance of the mixtures of E220 and E260 with Rosin resin at 50 % w/w.

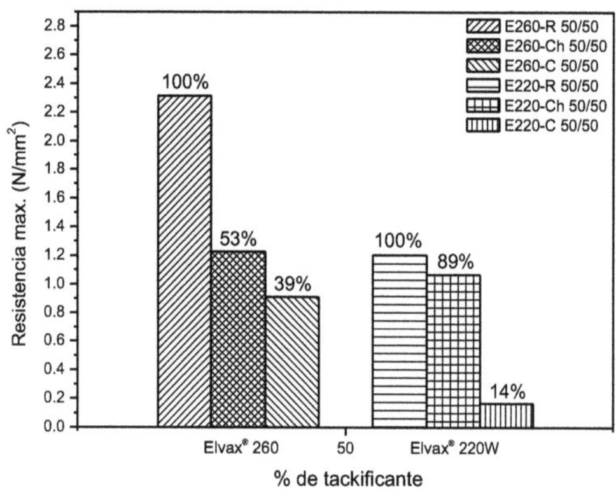

The maximum resistance values obtained in the tests carried out on the adhesives of E220 and E260 at 50% w/w decreased in the order of the tackifiers used: Rosin > G. *Chiloensis* > *G.* Camporum respectively (Figure 34).

Figure 34. Maximum resistance of the mixtures of E220 and E260 with Rosin resin, refined resin of G. *Chiloensis* and G. *Camporum* at 50%p/p.

The plasticizing effect of the tackifier in the adhesive mixture shows the decrease in resistance is more accentuated in the case of E260 than in that of E220. As can be seen in Figure 34 the resistance of E260-Rosin 50/50 and E220-G. *Chiloensis 50/50* are very similar indicating similar compatibility between E220 and Rosin and *G. Chiloensis* resin, and much higher than *G. Camporum resin*. It is also observed as in the case of Rosin resin, the mixtures of E220 and *Grindelia* have lower resistance than those formulated with E260, maintaining the relationship observed in Figure 32.

# Chapter IV

# Conclusions

# <u>CONCLUSIONS</u>

In the work presented, it was possible to determine and analyze the rheological, thermal and adhesive properties of the adhesives formulated with two ethylene vinyl acetate copolymers as base polymer and the natural resins of the genus *Grindelia* as tackifier, as well as the physical, rheological and thermal properties of its components separately.

By means of the Differential Scanning Calorimetry (DSC) technique, a good compatibility and miscibility of the studied systems defined by the existence of a single glass transition temperature (Tg) was found in all the EVA/Tackifier mixtures. The mixtures showed higher crystallization and melting values of ΔH than the values of the pure EVA copolymers, the latter indicating that the tackifiers may have acted as a nucleating agent of the crystalline part of both copolymers.

Flow properties showed that all consistency index values decreased with increasing temperature for both raw and refined Grindelia *Camporum* and *Chiloensis* resins. On the other hand, when analyzing the storage and loss modules (G' and G") it was determined that the crossing temperature decreased as the percentage of tackifier increased for all the studied cases, evidencing that the tackifier caused from the viscoelastic point of view a temperature shift in the passage from elastic solid to viscous liquid. On the other hand, the crossing temperatures were higher for the mixtures with E260 than for E220.

Shear strength properties decreased when reference blend tackifiers (EVA-Rosin) were replaced with *Grindelia* resins, however they can be used as replacements but for applications where shear properties are not as demanding.

# Chapter V

# Bibliography

# BIBLIOGRAPHY


- Aran Ais, F. (2000). Synthesis and characterization of thermoplastic polyurethanes containing rosin resins and their application as adhesives. Doctoral Thesis, University of Alicante, Spain.

- Barrueso - Martínez, M.L., Ferrándiz - Gómez, T.P., Cepeda - Jiménez, C.M., Sepulcre - Guilabert, J. y Martín - Martínez, J. M. (2001). Influence of vinyl acetate content and the tackifier nature on the rheological, thermal, and adhesión properties of EVA adhesives. *J. adhesión Science Technology*, 15, 243-263.

- Choi, W.Y., Lee, Ch.M. y Park, H.J. (2006). Development of biodegradable hot-melt adhesive based on poly-$\varepsilon$-caprolactone and soy protein isolate for food packaging system. *Food Science and Technology*, 39, 591-597.

- Dupont. (2005). Dupont Industrial Polymers: Thermal Properties of Elvax Measured by Differential Scanning Calorimeter (DSC). Dupont Elvax. *Product Information.* 1-4

- Dupont. (2005). For Adhesives, Sealants, and Wax Blends. Dupont Elvax. *Grade Selection Guide.* 1-7

- Dupont. (2009). Elvax resins Product Data Sheet. Elvax 220W. Dupont Packaging & Industrial Polymers. 1-3

- Dupont. (2009). Elvax resins Product Data Sheet. Elvax 260. Dupont Packaging & Industrial Polymers. 1-3

- Faker, M., Razavi Aghjeh M.K., Ghaffari, M. y Seyyedi, S.A. (2008). Rheology, morphology and mechanical properties of polyethylene/ethylene vinyl acetate copolymer (PE/EVA) blends. *European Polymer Journal*, 44, 1834-1842.

- Franck, A. J. (2005). Characterizing PSAs by rheology. SpecialChem. Adhesives and Sealants. httm://www.specialchem4adhesives.com

- Jackson, M. y Krajca, K. (2005). Going green. SpecialChem. Adhesives and Sealants. httm://www.specialchem4adhesives.com

- Li, Ch., Kong, Q., Zhao, J., Zhao, D., Fan, Q. y Xian, Y. (2004). Crystallization of partially miscible linear low-density polyethylene/poly(ethylene-co-vinylacetate) blends. *Materials Letters*, 58, 3613-3617.

- Martínez, L.M., De La Osa González, O. and Pollio, M.L. (2009). Study of the rheological properties of the resin extracted from the leaves and stems of *Grindelia Chiloensis*. *International Congress of Food Science and Technology*, Act I, 187.

- Petrie, E. M. (2005). Important characteristics of several common adhesives tests. SpecialChem. Adhesives and Sealants. httm://www.specialchem4adhesives.com

- Petrie, E. M. (2005). Thermal analysis techniques for adhesives. SpecialChem. Adhesives and Sealants. httm://www.specialchem4adhesives.com

- Przybytniak, G., Mirkowski, K. y Rafalski, A. (2007). Radiation degradation of blends polypropylene/poly(ethylene-co-vinyl acetate). *Radiation Physics and Chemistry*, 76, 1312-1317.

- Ramírez Guillen, A. (2005). Impact of the formulation on the properties of EVA-based hot melt adhesives containing hydrocarbon resins and waxes of different nature. PhD Thesis, University of Alicante, Spain.

- Ravetta, D. A., Anouti, A. y McLaughlin, S. P. (1996). Resin production of *Grindelia* accessions under cultivation. Industrial Crops and Products, 5, 197-201.

- Ravetta, D. A., Goffman, F., Pagano, E. y McLaughlin, S. P. (1996). *Grindelia chiloensis* resin and biomass production in its native Environment. Industrial Crops and Products, 5, 235-238.

- Roseberg, R.J. (1996). Underexploited temperate industrial and fiber crops, en J. Janick (Ed.). *Progress in new crops*, 60-84, ASHS Press, Alexandria, VA.

- Standard Test Method for Strength Properties of Adhesive Bonds in Shear by Compression Loading. Norma ASTM D 905 - 98.

- Wassner, D. F. y Ravetta, D. A. (2005). Temperature effects on leaf properties, resin content, and composition in *Grindelia chiloensis* (Asteraceae). Industrial Crops and Products, 21, 155-163.

- Wassner, D. F. y Ravetta, D. A. (2007). Nitrogen availability, growth, carbon partition and resin content in *Grindelia Chiloensis.* Industrial Crops and Products, 25, 218-230.

- Yoon, J-S., Oh, S-H, Kim, M-N, Chin, I-J y Kim, Y-H. (1999). Thermal and mechanical properties of poly(L-lactic acid)-poly(ethylene-co-vinyl acetate) blends. *Polymer*, 40, 2303-2312.

OMNIScriptum

Printed by Books on Demand GmbH, Norderstedt / Germany